Time

Peter Patilla

Heinemann Library
Des Plaines, Illinois

Designed by AMR
Illustrations by Art Construction and Jessica Stockam (Beehive Illustration)
Originated by HBM Print Ltd, Singapore
Printed and bound by South China Printing Co., Hong Kong/China

04 03 02 01 00
10 9 8 7 6 5 4 3 2

Library of Congress Cataloging-in-Publication Data
Patilla, Peter.
    Time / Peter Patilla.
       p. cm. –(Math links)
    Includes bibliographical references and index.
    Summary: Explores the measurement of time, including days, weeks,
months, and years, and introduces clocks and how to read them to
determine seconds, minutes, and hours.
    ISBN 1-57572-970-9 (lib. bdg.)
    1. Time measurements Juvenile literature. [1. Time measurements.
2. Time. 3. Clocks and watches.] I. Title. II. Series: Patilla,
Peter. Math links.
QB213.P38 1999
529—dc21   99-24956
                                 CIP

Acknowledgments
The Publishers would like to thank the following for permission to reproduce photographs:
Trevor Clifford, pp. 9, 11, 22, 23, 24, 25, 27, 29; Bruce Coleman Ltd./Kim Taylor, p. 7; Bruce Coleman
Ltd./Guido Cozzi, p. 13; Oxford Scientific Films/Warren Faidley, p. 5 top; Oxford Scientific Films/Stan
Osolinski, p. 10; Science Photo Library/Roger Harris, p. 18; Science Photo Library/Françoise Sauze, p.
21; Science and Society Picture Library, p. 20; Tony Stone Images /Lori Adamski Peek, p. 4; Tony Stone
Images/Olaf Soot, p. 5 bottom; Tony Stone Images/Zigy Kaluzny, p. 6 top; Tony Stone Images/
Richard Shock, p. 6 bottom; Tony Stone Images/David Hanover, p. 15; Tony Stone Images/Wayne
Eastep, p. 16, Tony Stone Images/George Robinson, p. 19.

Cover photo: Trevor Clifford

Our thanks to David Kirkby for his comments in the preparation of this book.

Every effort has been made to contact copyright holders of any material reproduced in this book.
Any omissions will be rectified in subsequent printings if notice is given to the Publisher.

Some words in this book are in bold, **like this**. You can find out
what they mean by looking in the glossary. Look for the answers
to questions in the green boxes on page 32.

Contents

A Day

A day is one way to measure time. It takes 24 hours for the earth to turn all the way around. Each day is 24 hours long. When it is night on one side of the earth, it is day on the other side.

At different times each day, the sun and moon appear in the sky. Sometimes you can see the moon while the sun is still shining.

At what time of day can you see the moon?

Day and Night

Nighttime is when it is dark. We usually sleep at night. Daytime is when it is light. We are usually awake during the day.

Not everything goes to sleep at night. Some animals are nocturnal. This means they are active at night.

Owls are nocturnal. They hunt at night.

Midnight and Midday

12 A.M. midnight

12 P.M. midday

Each new day starts at **midnight**. Half way through the day is **midday**. This is also called noon. Noon is when the sun is at its highest point in the sky.

Morning times have the same names as afternoon and evening times. The letters A.M. follow a time to show it is morning. The letters P.M. follow a time to show it is afternoon.

What time do you get up most days?

9

Morning

In the early morning, light breaks through the dark. This is called dawn. The sun rises over the **horizon**. This time is called sunrise.

Morning starts just after **midnight** and ends at noon. It is the time between 12 o'clock midnight and 12 o'clock noon.

Which meal do you eat in the morning?

Afternoon and Evening

Afternoon is the time between noon and evening. Many people think of afternoon as the time between lunch and supper.

The time between afternoon and night is called evening. It is the time between supper and bedtime. The time when the sun disappears below the **horizon** is called sunset.

What do you do most evenings?

A Week

Monday

Tuesday

Wednesday

Thursday

Friday

Saturday

Sunday

A week is seven days long. Some people start the week on a Monday. Other people start the week on a Sunday.

Saturday and Sunday are called the weekend.
The days Monday to Friday are often called
weekdays.

What do you do on weekends?

Months

There are twelve months in one year. The new year begins on January 1. The old year ends on December 31. People sometimes have fireworks at **midnight** on December 31 to celebrate the new year.

January

Sun	Mon	Tues	Wed	Thur	Fri	Sat
					1	2
3	4	5	6	7	8	9
10	11	12	13	14	15	16
17	18	19	20	21	22	23
24	25	26	27	28	29	30
31						

February

Sun	Mon	Tues	Wed	Thur	Fri	Sat
	1	2	3	4	5	6
7	8	9	10	11	12	13
14	15	16	17	18	19	20
21	22	23	24	25	26	27
28						

March

Sun	Mon	Tues	Wed	Thur	Fri	Sat
	1	2	3	4	5	6
7	8	9	10	11	12	13
14	15	16	17	18	19	20
21	22	23	24	25	26	27
28	29	30	31			

April

Sun	Mon	Tues	Wed	Thur	Fri	Sat
				1	2	3
4	5	6	7	8	9	10
11	12	13	14	15	16	17
18	19	20	21	22	23	24
25	26	27	28	29	30	

May

Sun	Mon	Tues	Wed	Thur	Fri	Sat
						1
2	3	4	5	6	7	8
9	10	11	12	13	14	15
16	17	18	19	20	21	22
23	24	25	26	27	28	29
30	31					

June

Sun	Mon	Tues	Wed	Thur	Fri	Sat
		1	2	3	4	5
6	7	8	9	10	11	12
13	14	15	16	17	18	19
20	21	22	23	24	25	26
27	28	29	30			

July

Sun	Mon	Tues	Wed	Thur	Fri	Sat
				1	2	3
4	5	6	7	8	9	10
11	12	13	14	15	16	17
18	19	20	21	22	23	24
25	26	27	28	29	30	31

August

Sun	Mon	Tues	Wed	Thur	Fri	Sat
1	2	3	4	5	6	7
8	9	10	11	12	13	14
15	16	17	18	19	20	21
22	23	24	25	26	27	28
29	30	31				

September

Sun	Mon	Tues	Wed	Thur	Fri	Sat
			1	2	3	4
5	6	7	8	9	10	11
12	13	14	15	16	17	18
19	20	21	22	23	24	25
26	27	28	29	30		

October

Sun	Mon	Tues	Wed	Thur	Fri	Sat
					1	2
3	4	5	6	7	8	9
10	11	12	13	14	15	16
17	18	19	20	21	22	23
24	25	26	27	28	29	30
31						

November

Sun	Mon	Tues	Wed	Thur	Fri	Sat
	1	2	3	4	5	6
7	8	9	10	11	12	13
14	15	16	17	18	19	20
21	22	23	24	25	26	27
28	29	30				

December

Sun	Mon	Tues	Wed	Thur	Fri	Sat
			1	2	3	4
5	6	7	8	9	10	11
12	13	14	15	16	17	18
19	20	21	22	23	24	25
26	27	28	29	30	31	

Most months have 30 or 31 days. A calendar shows the names of the months and the number of days in each month. It also shows the days of the week.

Look at the calendar. Which month does not have 30 or 31 days?

A Year

The earth takes just over 365 days to go around the sun. This is called a **year.** Most years have 365 days. A leap year has 366 days. There is a leap year once every four years.

A year is divided into twelve months. It is also divided into **seasons**. There are four seasons—spring, summer, fall, and winter. In some parts of the world, the weather changes with each season. Other parts of the world have wet and dry seasons.

What season does the picture show?

Measuring Time

Before there were clocks or watches, people used **hourglasses** to tell time. A larger hourglass measures a longer amount of time. A small hourglass measures a few minutes.

People also used **sundials** to tell the time. As the sun moves across the sky, a shadow on the face of the sundial moves. The shadow points to the correct time.

A sundial doesn't work at night or on a cloudy day. Why not?

Clocks and Watches

The **hands** on clocks and watches move. They circle the clock face to show the time. The shorter hand tells the hour. The longer hand tells the minutes.

second hand

hour hand

minute hand

The hands move at different speeds. The minute hand circles the face once each hour. The hour hand circles the face once every twelve hours.

Some clocks have another hand. It is called a second hand. It circles the face once each minute.

O'Clock Times

When the minute **hand** and the hour hand point to the number 12, it is 12 o'clock. The number the hour hand points to is the o'clock hour. O'clock is a short way of saying "of the clock."

Digital clocks or watches do not have hands. They show numbers. On a digital watch or clock, the o'clock time ends in two zeros. The first number tells you the o'clock time.

What o'clock times do these digital watches show?

Time and Quarter Hours

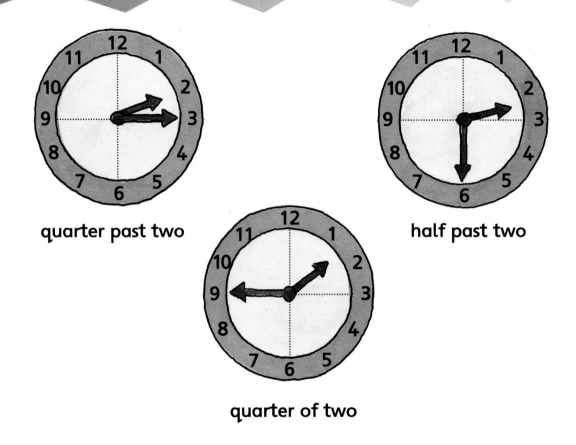

quarter past two

half past two

quarter of two

The clock face can be divided into halves and **quarters**. After the minute **hand** has gone one fourth around, we say it is quarter past the hour. Halfway around is half past the hour. For three fourths around, it is a quarter of the next hour.

It takes 15 minutes for the minute hand to move a quarter of the way around. Quarter past is the same as 15 minutes after the hour. Half past is the same as 30 minutes after the hour. Quarter of is the same as 15 minutes before the next hour.

What times do these clocks show?

Minutes Past Time

It takes 5 minutes for the minute **hand** to move
from one number to the next. It takes 60 minutes
for the minute hand to go all the way around.
Some watches and clocks have the minutes
written on the face.

To write the time, we write the hour first. Then we write the number of the minutes past the hour. So 9:20 is 20 minutes past 9 o'clock. **Digital** watches show time the same way we write it.

What times do these watches show?

29

Glossary

A.M. any time between midnight and noon. The letters are short for the Latin words *ante meridian,* meaning before the sun is at its highest point.

digital clocks and watches that show the time with numbers

half past time when the minute hand is halfway around the clock face and points to the number 6

hand pointer on a clock or watch that shows the time

horizon imaginary line where the land or water meets sky

hourglass device that measures time by letting sand pour slowly from one glass bulb to another

midday 12 o'clock in the middle of the day; also called noon

midnight 12 o'clock in the middle of the night–the end of one day and the start of the next

P.M. any time between noon and midnight. The letters are short for the Latin words *post meridian,* meaning after the sun has been at its highest point.

quarter one of the equal parts of something that has been divided into four parts

season one of four parts of the year

sundial instrument that tells the time by the position of the sun's shadow on a dial

year time it takes the earth to go around the sun, about 365 days

Fact File

Time facts

1 minute = 60 seconds
15 minutes = 1 quarter hour
30 minutes = 1 half hour
60 minutes = 1 hour
24 hours = 1 day

7 days = 1 week
30 or 31 days = 1 month
12 months = 1 year
365 days = 1 year

Year facts

- Months with 31 days: January, March, May, July, August, October, December
- Months with 30 days: April, June, September, November
- February usually has 28 days. Every four years February has 29 days. This is called a leap year. Leap years have 366 days.

More Books to Read

Chapman, Gillian. *Exploring Time*. Danbury, Conn.: Millbrook Press, 1995.

Ganeri, Anita. *The Story of Time & Clocks*. New York: Oxford University Press, 1996. An older reader can help you with this book.

Hewitt, Sally. *Time*. Austin, Tex.: Raintree Steck-Vaughn, 1996.

Answers

page 5   evening and night
page 11   breakfast
page 17   February
page 19   fall
page 21   There is no sunlight to cast a shadow.
page 25   (top) 7:00 7 o'clock
                 (bottom) 12:00 12 o'clock
page 27   1) a quarter past 10 (10:15)
                 2) half past 2 (2:30)
                 3) a quarter to 11 (10:45)
page 29   1) 9:20
                 2) 9:25
                 3) 9:30
                 3) 9:35

Index